Collins

AQA GCSE 9-1
Maths
Foundation

Practice Papers

Mike Fawcett and Keith Gordon

Contents

SET A

SET B

ANSWERS

Acknowledgements

The authors and publisher are grateful to the copyright holders for permission to use quoted materials and images.

All images are © HarperCollinsPublishers and Shutterstock.com

Every effort has been made to trace copyright holders and obtain their permission for the use of copyright material. The author and publisher will gladly receive information enabling them to rectify any error or omission in subsequent editions. All facts are correct at time of going to press.

Published by Collins
An imprint of HarperCollinsPublishers
1 London Bridge Street
London SE1 9GF

HarperCollinsPublishers
Macken House, 39/40 Mayor Street Upper,
Dublin 1, D01 C9W8

© HarperCollinsPublishers Limited 2020

ISBN 9780008321383

First published 2019

This edition published 2023

10 9 8 7 6 5

All rights reserved. No part of this publication may be reproduced, stored in a retrieval system, or transmitted, in any form or by any means, electronic, mechanical, photocopying, recording or otherwise, without the prior permission of Collins.

British Library Cataloguing in Publication Data.

A CIP record of this book is available from the British Library.

Commissioning Editor: Kerry Ferguson
Project Leader and Management: Chantal Addy and Richard Toms
Authors: Mike Fawcett and Keith Gordon
Consultant Author: Trevor Senior
Cover Design: Sarah Duxbury and Kevin Robbins
Inside Concept Design: Ian Wrigley
Text Design and Layout: QBS Learning
Production: Karen Nulty
Printed by Ashford Colour Press Ltd

Collins

AQA
GCSE
Mathematics

F

SET A – Paper 1 Foundation Tier

Author: Mike Fawcett

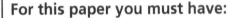

Materials

Time allowed: 1 hour 30 minutes

For this paper you must have:

- mathematical instruments

You may **not** use a calculator.

Instructions

- Use black ink or black ball-point pen. Draw diagrams in pencil.
- Answer **all** questions.
- You must answer the questions in the space provided.
- In all calculations, show clearly how you work out your answer.

Information

- The marks for questions are shown in brackets.
- The maximum mark for this paper is 80.
- You may use additional paper, graph paper and tracing paper.

Name: _____

Answer **all** questions in the spaces provided.

1 **(a)** Write down the factors of 8

[2 marks]

Answer ....

1 **(b)** Write down **two** multiples of 11

[1 mark]

Answer ....

2 Write down the value of 4^2

[1 mark]

Answer ....

3 Expand $2(x + 4)$

[1 mark]

Answer ....

4

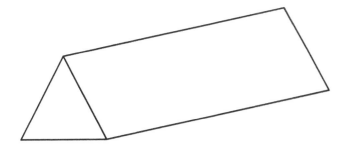

4   (a) Give the mathematical name for this 3-D shape.   [1 mark]

     Answer _____

4   (b) Which statements are correct for this shape?

                                                                           [1 mark]

     Tick **two** boxes.

          It has 6 faces   ☐

          It has 6 vertices   ☐

          It has 6 edges   ☐

          It has 9 edges   ☐

          It has 5 vertices   ☐

5   Convert $2\frac{3}{4}$ to an improper fraction.

                                                                           [1 mark]

     _____

                                        Answer _____

6   Write   £1.25 to 75p   as a ratio in its simplest form.

                                                                           [2 marks]

     _____

     _____

                      Answer _____ _____

7 **(a)** Write down the next term in the sequence.

1   1   2   3   5   8

[1 mark]

Answer _____

7 **(b)** A pattern is made using matchsticks.

Pattern 1   Pattern 2   Pattern 3

How many matchsticks will be needed to make the 50th pattern?

[3 marks]

_____

_____

_____

_____

Answer _____

8   Molly is thinking of a number.

She squares it and then adds 15

She gets an answer of 64

What number was Molly thinking of?

[2 marks]

Answer

9   Tennis balls cost 48p each or packs of 3 cost £1.25

What is the smallest possible cost for 40 tennis balls?

[4 marks]

Answer £

10 468 students were asked to pick an activity.

The pie chart shows the activities that the students chose.

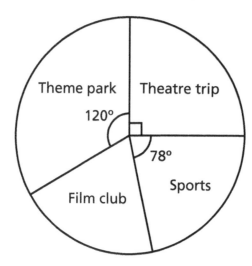

10 **(a)** How many students chose the theatre trip?

[2 marks]

Answer ....

10 **(b)** What percentage of students chose the film club?

[3 marks]

Answer .... %

11 **(a)** Solve $\dfrac{x}{5} = 3$

[1 mark]

Answer ....

11 **(b)** Solve $3x - 5 \leqslant 4$

[2 marks]

Answer ....

11 **(c)** Simplify $\dfrac{12x^5}{3x^2}$

[2 marks]

Answer ....

12 Jessica has uploaded 10 videos online.

Here is the number of views for each video.

181   204   175   153   97   11   113   169   165   198

12 **(a)** Pete says, "The range is 198 – 181 = 17"

Explain his mistake.

[1 mark]

....

....

12 **(b)** Explain why the median would be a more suitable average than the mean for this data.

[1 mark]

....

....

12 **(c)** Calculate the median number of views for Jessica's videos.

[3 marks]

....

....

Answer: Median = ...

13 The diagram shows three aeroplanes A, B and C near to an airport tower, T.

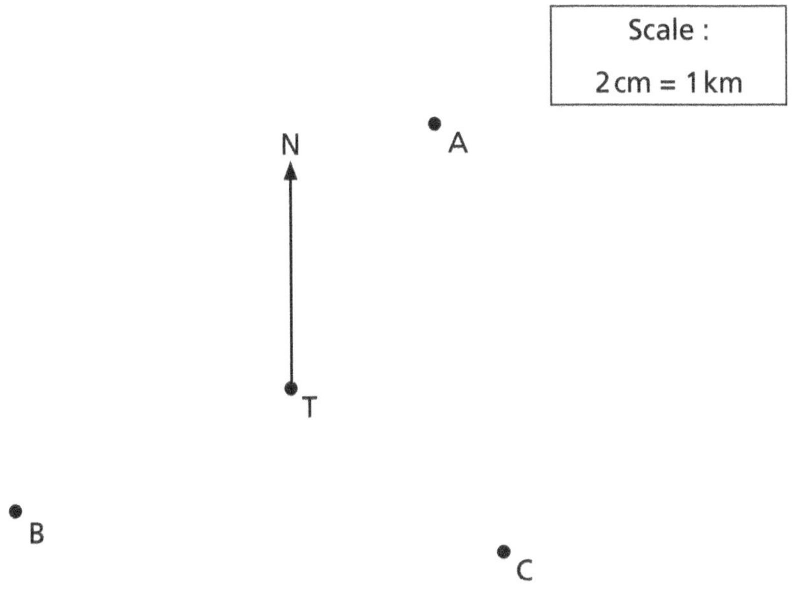

Scale :
2 cm = 1 km

N

• A

• T

• B

• C

13 **(a)** Work out the bearing of aeroplane C from the tower.

[1 mark]

Answer _____ °

13 **(b)** How far is aeroplane B from the tower?

[1 mark]

Answer _____ km

13 **(c)** A fourth aeroplane, D, is on a bearing of 290° from the tower.

It is 2.2 km from the tower.

Mark this position on the diagram.
Label it D.

[2 marks]

14 Rachel is at the gym for 2 hours.

She spends $\frac{2}{5}$ of her time on the weights.

The rest of her time is spent running and cycling in the ratio of 4 : 5

How many minutes does she spend cycling?

[4 marks]

Answer .... minutes

15 A machine can make 53 items in 5.8 minutes.

15 **(a)** Estimate the number of items the machine can make in one day.

[2 marks]

Answer ....

15 **(b)** State any assumptions that you have made.

[1 mark]

16 Ethan, Bob and Josh throw a bottle and try to land it upright.

Here are the results.

	Ethan	Bob	Josh
Number of tries	10	25	50
Number of lands upright	1	3	4

16 (a) Who is the best at the game?

Give a reason to support your answer.

[2 marks]

16 (b) Whose results give you a better understanding of their ability?

Give a reason for your decision.

[2 marks]

17 (a) Work out $\dfrac{2}{5} + \dfrac{1}{3}$

[2 marks]

Answer

17 (b) Work out $\dfrac{9}{2} \div 6$

[2 marks]

Answer

18   Tim cycles along a road to test his new bike.

He stops on the way to adjust his brakes.

The graph shows his journey along the road.

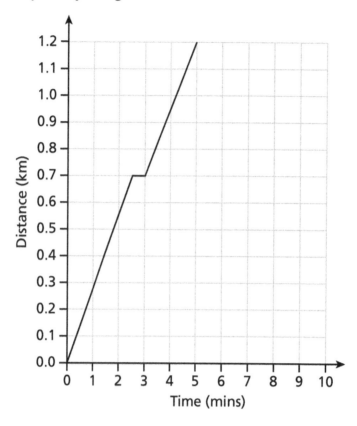

18   **(a)**   How long did he spend adjusting his brakes?

[1 mark]

Answer ....

18   **(b)**   Work out his average speed from home to the end of the road.

Give your answer in metres per second (m/s).

[2 marks]

....

....

....

Answer .... m/s

18 **(c)** Tim rests for 1 minute at the end of the road before cycling back.
He then cycles back at 6 m/s.

Complete the graph for his journey.

[3 marks]

19 Ali, Brad and Dora each have some marbles.

Brad has 4 times as many as Ali.

Dora has 12 more than Ali.

Together, Ali and Brad have the same number of marbles as Dora.

How many marbles does Ali have?

[3 marks]

Answer

20 The exact volume of the cylinder is 320π cm³

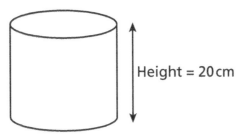

Height = 20 cm

Work out the diameter of the cylinder.

[4 marks]

Answer cm

21   The table shows the population of some countries, given in standard form.

Country	Population
Japan	1.2×10^8
France	6.5×10^7
China	1.4×10^9
UK	6.8×10^7
US	3.4×10^8
Egypt	1.1×10^8

21   (a) Write the population of Egypt as an ordinary number.

[1 mark]

Answer ....

21   (b) List the countries in order of their population size from smallest to largest.

[2 marks]

....

....

Answer ....

21   (c) The populations have been rounded to 2 significant figures.

Write down the error interval for the actual population of the UK.

[2 marks]

Answer .... ⩽ Population of UK < ....

22 The running time of a new feature film is reduced by 20%.

The film is now 100 minutes long.

How long was the original film?

[2 marks]

Answer _____ minutes

23 **(a)** Write down the exact value of   sin 30°

[1 mark]

sin 30° = _____

23 **(b)** Work out the length of the side marked x.

[3 marks]

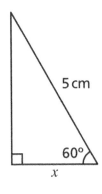

Not drawn accurately

5 cm

60°

x

$x =$ _____ cm

24   The probability that Leah will win her tennis match is 5^{-1}

Work out the probability that she will **not** win.

Give your answer in the form $\dfrac{a}{b}$

[2 marks]

Answer ....

END OF QUESTIONS

   ©HarperCollins*Publishers* 2019

Collins

AQA
GCSE

Mathematics

F

SET A – Paper 2 Foundation Tier

Author: Mike Fawcett

Materials

Time allowed: 1 hour 30 minutes

For this paper you must have:

- calculator
- mathematical instruments

Instructions

- Use black ink or black ball-point pen. Draw diagrams in pencil.
- Answer **all** questions.
- You must answer the questions in the space provided.
- In all calculations, show clearly how you work out your answer.

Information

- The marks for questions are shown in brackets.
- The maximum mark for this paper is 80.
- You may use additional paper, graph paper and tracing paper.

Name: _____

Answer **all** questions in the spaces provided.

1   Round   2.359   to 1 decimal place.

                                   [1 mark]

Answer _____

2   **(a)** Convert   2.34 kg   to grams.

                                   [1 mark]

Answer _____ grams

2   **(b)** Convert   6.4 m   to centimetres.

                                   [1 mark]

Answer _____ cm

3   Simplify $3x + 4y - x - 6y$

                                   [2 marks]

Answer _____

4   Here is a rectangle.

$3a$ cm

4 cm

4   **(a)**   Write down an expression for the area of the rectangle.

[1 mark]

Answer _____ cm^2

4   **(b)**   The area of the rectangle is 60 cm^2

Work out the perimeter of the rectangle.

[3 marks]

Answer _____ cm

5 In total, Tim's chickens laid 27 eggs this week.

The pictogram shows the number of eggs Tim collected from Monday to Friday.

Day	Frequency
Monday	⬭⬭
Tuesday	⬭⬭⬭
Wednesday	⬭◖
Thursday	◖
Friday	⬭⬭⬭◖

Key: ⬭ = 2 eggs

Tim collected the same number of eggs on Saturday and Sunday.

How many did he collect on Sunday?

[2 marks]

....

....

Answer ....

6 Work out the coordinates of the midpoint of *AB*.

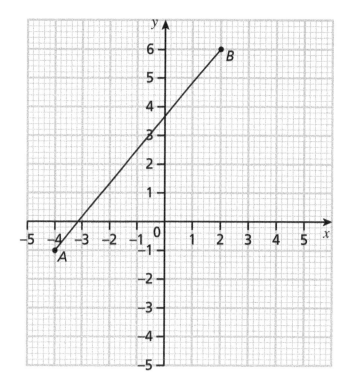

[2 marks]

....

....

Answer (.................... , ...)

7   Yusuf earns £36 per week.

7   **(a)**   One week, he spends $\frac{3}{5}$ of it and saves the rest.

How much money does he save?

[2 marks]

Answer £

7   **(b)**   Another week, he saves £16

What percentage of his money did he spend that week?

[3 marks]

Answer   %

8   (a) Doug is going to measure some students in order to compare the heights of boys and girls.

Which words describe his data?

Tick **two** boxes.

[1 mark]

Primary ☐

Secondary ☐

Discrete ☐

Continuous ☐

8   (b) The table summarises his results.

	Median height	Range of heights
Boys	1.52 m	0.36 m
Girls	1.47 m	0.42 m

Use his results to make **two** comparisons between the heights of boys and girls.

[2 marks]

9   A teacher writes this calculation $2 + 3 \times 4 - 8$

9   **(a)** Ikra says that the answer is 12

Is she correct?
You **must** show your working.

[2 marks]

9   **(b)** Hannah says, "I can make the answer equal –10, just by including brackets."

Show that Hannah is correct.

[1 mark]

10   Given that $2250 = 2 \times 3^a \times 5^b$

Work out the values of a and b.

[2 marks]

$a =$ ....

$b =$ ....

11 50 vegetarians were asked if they eat eggs or dairy products.

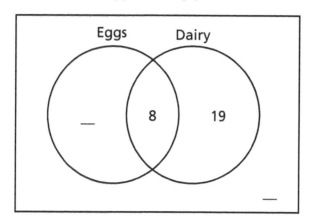

11 **(a)** 21 said that they eat eggs.

Complete the Venn diagram.

[2 marks]

11 **(b)** One of the vegetarians is chosen at random.

What is the probability that they eat neither eggs nor dairy?

[2 marks]

Answer ....

12 Write the following in order from smallest to largest.

$\dfrac{3}{7}$ 0.41   38.5%

[2 marks]

Answer ....

13   Sean, Nicole and Jadyn are taking part in a long jump competition.

Jadyn can jump $\frac{3}{5}$ of the distance that Sean can jump.

Nicole can jump twice as far as Jadyn.

Write the distances that Jadyn, Sean and Nicole can jump as a **ratio** in its simplest form.

[3 marks]

Answer _____ : _____ : _____

14   In the diagram, the numbers opposite each other are factor pairs of the number in the centre.

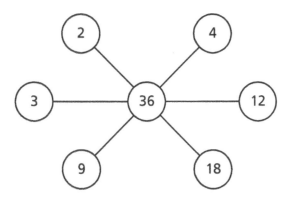

Write numbers in the diagram below, so that it follows the same rules.
You must not repeat any factor pairs.

[3 marks]

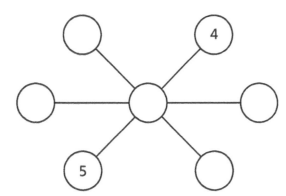

15 Convert 2 m² into cm² [2 marks]

Answer _____ cm²

16 **(a)** Factorise fully $4x^2 - 12x$ [2 marks]

Answer _____

16 **(b)** Find the largest integer for x that satisfies $7x - 18 < 3x$ [2 marks]

Answer _____

17 The size of angle x to the size of angle y = 2 : 1

Not drawn accurately

17 **(a)** Show that angle x = 72 degrees [3 marks]

17 **(b)** What is the size of angle z?

Give a reason for your answer. **[2 marks]**

$z =$ .... degrees

18 A helicopter flies in a straight line so that it is an equal distance from the lighthouse and the cliffs.

When it is exactly 500 m from the yacht, it hovers in a fixed position.

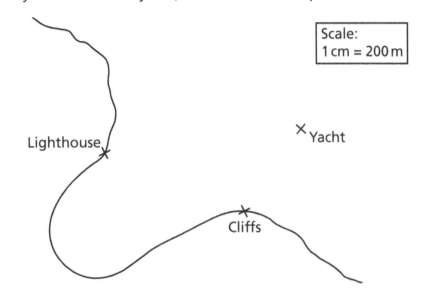

Scale:
1 cm = 200 m

× Yacht

Lighthouse

Cliffs

Mark, with a cross, the position of the helicopter as it hovers. **[3 marks]**

19 You are given that Pressure = $\dfrac{\text{Force}}{\text{Area}}$

The bookcase exerts a downward force of 3000 N.

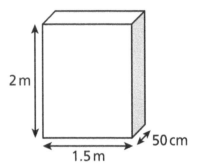

2 m

1.5 m

50 cm

Use the formula to work out the pressure applied to the base of the bookcase. **[3 marks]**

....

....

....

Answer .... N/m²

20   The amount of fencing that a company produces each day is proportional to the number of workers.

The table shows information for four days.

	Mon	Tue	Wed	Thu
Number of workers	3	5	8	5
Length of fencing produced	645 cm	10.75 m	17.2 m	10.75 m

On Friday, there are 7 workers.

Work out the number of metres of fencing the company will produce.

[3 marks]

Answer .... m

21   Work out the area of the semicircle.

Circle the correct answer.

[1 mark]

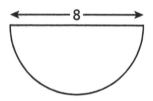

4π   8π   16π   32π

22 In a café, to make a coffee

it takes Bel 42 seconds

and it takes Iris 70 seconds.

Bel says, "I can make x coffees while you are only making y coffees."

Work out the least possible values of x and y to make the statement correct.

[3 marks]

$x =$ ....

$y =$ ....

23 Translate shape A with the vector $\begin{pmatrix} -2 \\ -4 \end{pmatrix}$ and label the new shape B.

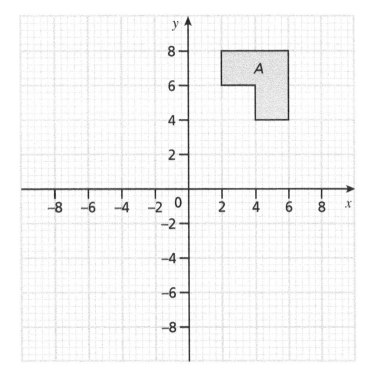

[2 marks]

24 A house is worth £159 000

This is a 6% increase on its value last year.

Work out the value last year.

[3 marks]

Answer £ _____

25 The table summarises the heights of 20 students.

Height, h cm	Frequency	Mid value	
$140 < h \leqslant 150$	3		
$150 < h \leqslant 160$	6		
$160 < h \leqslant 170$	7		
$170 < h \leqslant 180$	4		

Work out an estimate of the mean height.

[4 marks]

Answer _____ cm

26 **(a)** By plotting the graph of $3y = 5x + 3$, solve the simultaneous equations

$$3y = 5x + 3$$
$$x + y = 5$$

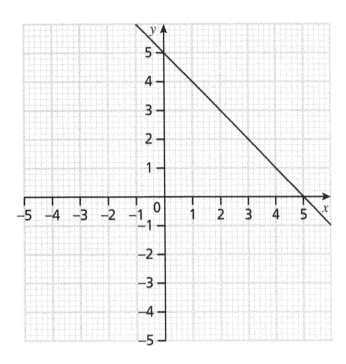

[4 marks]

$x = $ ....

$y = $ ....

26 **(b)** Work out the equation of the line which is parallel to $x + y = 5$ and goes through the point (3, 4).

[2 marks]

....

....

....

Answer ....

27 The graph shows the height of water in a container which is left out overnight in the rain.

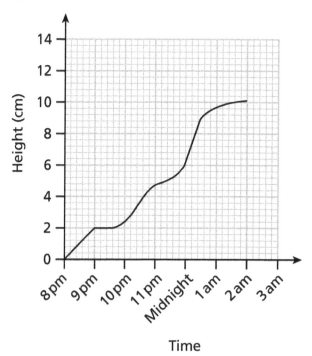

Time

27 **(a)** At what time does it stop raining?

[1 mark]

Answer ....

27 **(b)** What can you say about the rainfall between 8 pm and 9 pm?

[1 mark]

....

27 **(c)** At 2 am, it continues to rain.

The height of water in the container increases by another 3 cm in the next hour.

Show this on the graph.

[1 mark]

END OF QUESTIONS

Collins

AQA
GCSE
Mathematics
SET A – Paper 3 Foundation Tier
Author: Mike Fawcett

F

Materials

Time allowed: 1 hour 30 minutes

For this paper you must have:
- calculator
- mathematical instruments

Instructions

- Use black ink or black ball-point pen. Draw diagrams in pencil.
- Answer **all** questions.
- You must answer the questions in the space provided.
- In all calculations, show clearly how you work out your answer.

Information

- The marks for questions are shown in brackets.
- The maximum mark for this paper is 80.
- You may use additional paper, graph paper and tracing paper.

Name: _____

Answer **all** questions in the spaces provided.

1 Write down the value of the digit 9 in the number 231.92   **[1 mark]**

Answer ....

2 Work out   −6 + (−9)   **[1 mark]**

....

Answer ....

3

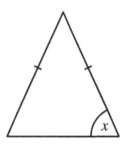

3 **(a)** Give the name for this special triangle.   **[1 mark]**

Answer ....

3 **(b)** Measure the size of angle x.   **[1 mark]**

Answer .... degrees

4 Write these numbers in ascending order.

       1.3   1.03   1.33   1.303

[2 marks]

....

....

Answer ....

5 36 people were asked to choose an activity during a health and well-being evening.

The bar chart shows the results.

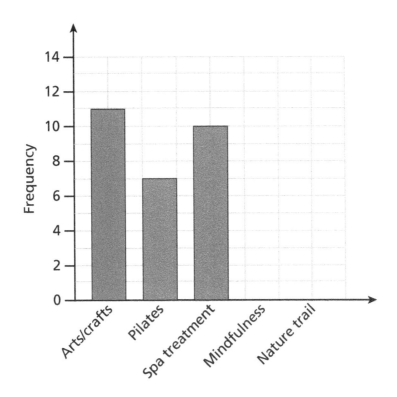

The same number of people chose mindfulness and nature trail.

Complete the bar chart.

[3 marks]

6 **(a)** Write down the next term in the sequence.

3   6   12   24   …

[1 mark]

Answer _____

6 **(b)** Is the number 140 a term in this sequence?
Give a reason for your answer.

[1 mark]

_____

_____

7   Points *A* and *B* are plotted on the coordinate grid.

ABCD is a rectangle on the grid with an area of 21 cm²

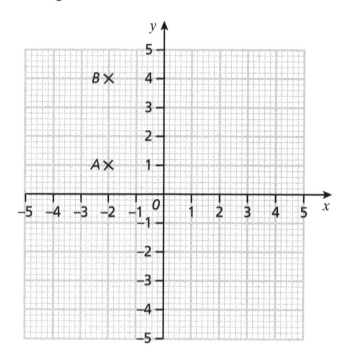

7   **(a)** Write down the coordinates of point *B*.

[1 mark]

Answer (............,............)

7   **(b)** Work out the coordinates of points *C* and *D*.

[2 marks]

Answer *C* = (............,............)

Answer *D* = (............,............)

7   **(c)** Work out the perimeter of the rectangle *ABCD*.

[2 marks]

....

....

Answer .... cm

8   A boat starts at the harbour (H); it travels **due north** for 2 miles, then **due west** for 2 miles.

In which compass direction will the boat need to travel to go directly back to the harbour?

[2 marks]

N

x H

Answer ....

9   (a) Calculate the square root of   2.25

[1 mark]

Answer ....

9   (b) Calculate the cube of   2.1

[1 mark]

Answer ....

9   (c) Calculate 4^5

[1 mark]

Answer ....

10   Jill says, "My house number is

a factor of 40

a cube number

not a square number."

What is the house number?

[3 marks]

Answer

11   A tank contains 25 000 litres of water.

Water leaks from the tank at a rate of 84 litres per day.

After how many **whole** days is the tank less than half full?

[4 marks]

Answer .... days

12   This flow diagram can help to solve the equation $y = x^2 - 5$

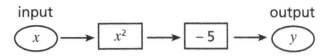

input   output

x → x^2 → -5 → y

12   **(a)**   Work out the value of y when $x = 6$

[1 mark]

$y =$ ....

12   **(b)**   Work out the value of x when $y = 59$

[1 mark]

$x =$ ....

13   Decide whether each of the following statements are **true**, **sometimes true** or **false**.

[4 marks]

$2 < -3$ Answer ....

$-1 > -3$ Answer ....

$x^2 = x$ Answer ....

$x^2 \geqslant 0$ Answer ....

14   Ed prints metallic sign in three sizes.

The tables show the prices he charges and his costs for printing and postage.

Size	Price
A5	£5.95
A4	£8.65
A3	£10.85

Size	Printing costs
A5	£1.07
A4	£1.52
A3	£3.09

Number of items per customer	Postal costs
1 sign	£2.40
2 or more signs	£3.80

Ed has this offer

- 3 for the price of 2
- free delivery

He makes these sales to three customers.

- Two A3 signs and one A4 sign
- Two A5 signs
- One A4 sign

How much profit did Ed make from the three customers?

[6 marks]

Answer £ _____

15 Which calculation works out the simple interest earned on an investment of £2500 at 3% for 2 years?

Circle the correct answer. [1 mark]

£2500 × 3 × 2   £2500 × 0.3 × 2   £2500 × 0.03 × 2   £2500 × 1.03 × 2

16 A sports centre has a gym and a swimming pool.

On Wednesday, 51 people visited the centre.

13 people who used the gym also went swimming.

21 people did **not** use the gym and 6 of those did **not** swim either.

Complete the frequency tree to show this information.

[2 marks]

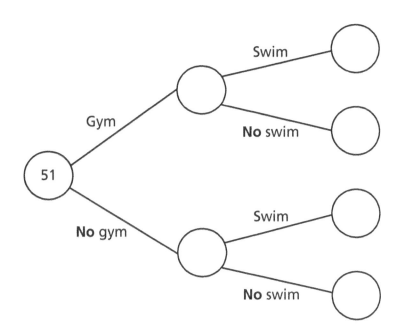

17 A total of 584 men, women and children are in a shopping centre.

312 are women and approximately 18.8% are children.

17 **(a)** What fraction of the people in the shopping centre are men?

Give your answer in its simplest form.

[3 marks]

Answer

17 **(b)** The table shows the number of people who have left and the number of people who have arrived at the shopping centre.

	Men	Women	Children
Left	10	12	7
Arrived	11	30	6

What percentage of the people in the shopping centre now are women?

[3 marks]

Answer _____ %

18 A metal rod for a piece of machinery is $3\frac{4}{5}$ inches long.

The designer says that the length of the rod needs to be increased by one-third.

How long should the rod be?

Give your answer as a mixed number. **[3 marks]**

Answer .... inches

19 10 people were asked their height and annual income.

The table shows the results.

Income (£)	14 000	21 000	26 500	32 500	28 500	15 000	13 000	25 000	33 500	29 000
Height (m)	1.59	1.72	1.85	1.65	1.57	1.83	1.71	1.65	1.79	1.72

19 **(a)** Plot a scatter graph for this data. **[1 mark]**

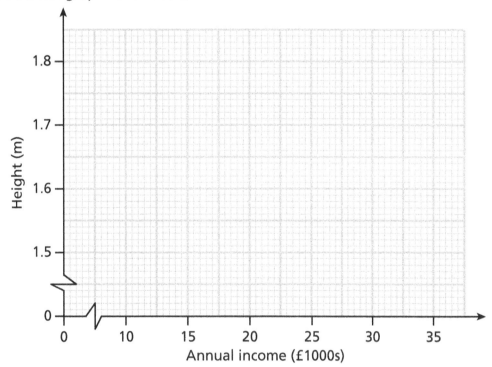

19 **(b)** Describe the correlation and state how annual income is related to height. **[2 marks]**

20 Work out $(7.2 \times 10^{12}) \div (1.5 \times 10^4)$

Give your answer in standard form.

[2 marks]

Answer ....

21 Here is a conversion graph to change pounds (£) to euros (€).

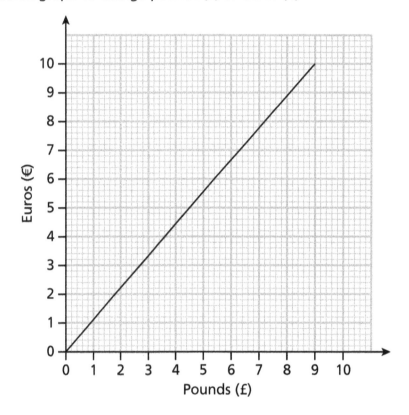

John is flying to Paris and then to Beirut.

He changes £500 into euros to take to Paris.

21 **(a)** Convert £500 to euros.

[1 mark]

Answer € ....

21 **(b)** John spends €260 in Paris and changes the rest into Lebanese Pounds (LBP).

The exchange rate is £1 = 1990 LBP

How many Lebanese Pounds does he take to Beirut?

[3 marks]

Answer _____ LBP

22 The interior angle of a regular polygon is 150 degrees.

Show that the polygon has 12 sides.

150°

[3 marks]

23 A rectangle has length 8 cm and width 5 cm.

A similar rectangle has length 12 cm.

Work out its width.

[3 marks]

Answer _____ cm

24 Volume of a cone $= \frac{1}{3}\pi r^2 h$

The cone has base radius 3 cm and height 5 cm.

24 **(a)** Work out the volume of the cone.

[2 marks]

....

....

Answer .... cm³

24 **(b)** The mass of the cone is 336 grams.

Work out its density.

[2 marks]

....

....

Answer .... g/cm³

25 **(a)** Factorise $\quad x^2 + x - 6$

[2 marks]

Answer

25 **(b)** Plot the graph of $\quad y = x^2 + x - 6$

[3 marks]

x	−3	−2	−1	0	1	2	3
y		−4			−4	0	6

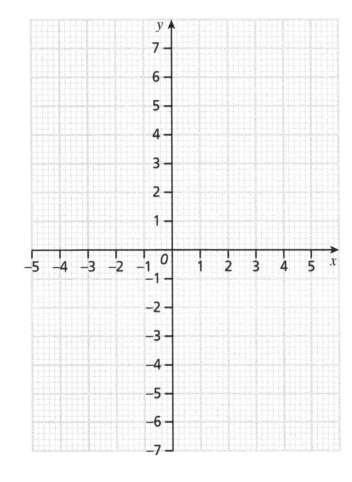

26 The probability that a biased coin lands on heads is 0.8

The coin is thrown twice.

26 **(a)** Complete the tree diagram.

[2 marks]

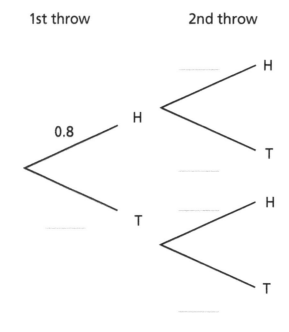

1st throw   2nd throw

26 **(b)** Work out the probability of throwing two heads.

[2 marks]

Answer _____

END OF QUESTIONS

Collins

AQA
GCSE
Mathematics

F

SET B – Paper 1 Foundation Tier

Author: Keith Gordon

Materials

Time allowed: 1 hour 30 minutes

For this paper you must have:

- mathematical instruments

You may **not** use a calculator.

Instructions

- Use black ink or black ball-point pen. Draw diagrams in pencil.
- Answer **all** questions.
- You must answer the questions in the space provided.
- In all calculations, show clearly how you work out your answer.

Information

- The marks for questions are shown in brackets.
- The maximum mark for this paper is 80.
- You may use additional paper, graph paper and tracing paper.

Name: _____

Answer **all** questions in the spaces provided.

1   Convert   3.5 kilometres   to metres.

[1 mark]

Answer _____ m

2   Here are five numbers.

8   9   5   7   2

2   **(a)** Write down the range.

[1 mark]

Answer _____

2   **(b)** Work out the median.

[1 mark]

Answer _____

3   Write down a fraction that is between $\frac{1}{3}$ and $\frac{3}{5}$

[1 mark]

Answer _____

4   40 people are asked to comment on the service in a restaurant.

The **pictogram** shows some of the results.

Excellent	○ ○ ○ ○ ◔
Very good	○ ○ ◕
Average	○ ◗
Poor	○
Very poor	

17 people said the service was excellent.

4   **(a)** Complete the key below.

[1 mark]

○ represents _____ people

4   **(b)** How many people said the service was very good?

[1 mark]

Answer _____

4   **(c)** How many people said the service was average or better?

[2 marks]

Answer _____

4   **(d)** Complete the pictogram.

[2 marks]

5   (a) Work out   736 + 249

[1 mark]

....

Answer ....

5   (b) Work out   323 − 156

[1 mark]

....

Answer ....

5   (c) Work out   6 × 23

[1 mark]

....

Answer ....

5   (d) Work out   128 ÷ 4

[1 mark]

....

Answer ....

6   In a game a prize is hidden in one of 12 boxes.

| 1 | 2 | 3 | 4 | 5 | 6 | 7 | 8 | 9 | 10 | 11 | 12 |

Mia is playing the game.

She is told that the prize is:

not in a box that is a multiple of 3

in a box that is a prime number

nearer to box 1 than box 12.

Which boxes could the prize be in?

[2 marks]

....

....

Answer ....

7   Here are two train timetables.

Denby Dale	0624	0724	0824	0924	1024
Huddersfield	0652	0752	0852	0952	1052

Huddersfield	0702	0802	0835	0916	1002
Manchester Airport	0750	0850	0925	1005	1050

To travel from Denby Dale to Manchester Airport, you change trains at Huddersfield.

7   **(a)**   Mary is catching a train from Denby Dale to Manchester Airport.

Her plane is due to depart at 1230

She has to arrive at the airport 3 hours before the plane departs.

What is the time of the latest train she can catch from Denby Dale?   **[1 mark]**

Answer ....

7   **(b)**   Arthur catches the 0824 from Denby Dale.

How long is his journey to the airport?

Assume he catches the earliest train possible from Huddersfield.   **[3 marks]**

....

....

Answer ....

7   **(c)**   Zak is at Huddersfield Station.

He looks at his watch.

How long will he have to wait for the next train to Manchester Airport?   **[2 marks]**

....

....

Answer .... minutes

8   Eggs are delivered in trays containing 24 eggs.

A hotel orders 32 trays.

How many eggs does the hotel order?

[3 marks]

Answer ....

9 $A(1, 2)$, $B(2, 6)$, $C(8, 6)$ and $D(7, 2)$ are the four vertices of a quadrilateral.

9   **(a)**   Draw the quadrilateral on the centimetre grid.

[2 marks]

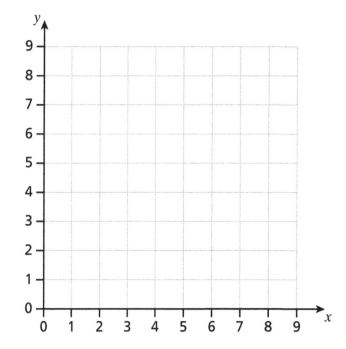

9 **(b)** What type of quadrilateral is *ABCD*?

[1 mark]

Answer ....

9 **(c)** Work out the area of *ABCD*.

[2 marks]

....

....

Answer .... cm^2

10 **(a)** Simplify $7a + 6a - 5a$

[1 mark]

....

Answer ....

10 **(b)** Simplify fully $2 \times 3m + 6 \times 5m$

[2 marks]

....

Answer ....

11   The conversion graph compares acres to hectares.

Acres are a measurement of area that is commonly used in Britain.

Hectares are a metric unit of area.

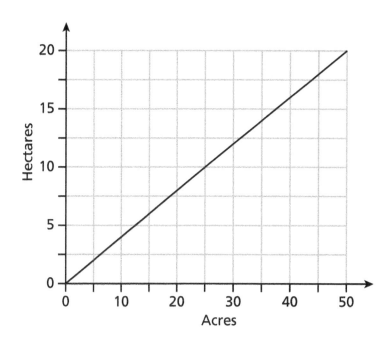

11   **(a)**   How many acres are there in 15 hectares?

[1 mark]

Answer ....

11   **(b)**   100 acres of farmland is for sale.

Farmland has a value of £25 000 per hectare.

Approximately, what is the value of the farmland?

[3 marks]

....

....

....

....

Answer £ ....

12 56 men and 66 women are asked if they can swim.

$\frac{4}{7}$ of the men say yes

$\frac{9}{11}$ of the women say yes

How many of the people asked can swim?

[3 marks]

Answer _____

13 **(a)** Here is a fair spinner.

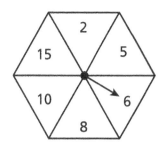

On the probability scale show the probability that the spinner lands on an odd number.

[1 mark]

0   1

13 **(b)** On this fair spinner, write numbers in each sector so that

the probability of the arrow landing on an odd number is $\frac{1}{2}$

the probability of the arrow landing on a multiple of 3 is $\frac{1}{3}$

[2 marks]

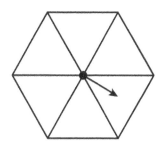

14 *ABC* and *ACD* are triangles.

AC = CD = AB

BCD is a straight line.

Angle *BAC* = 20°

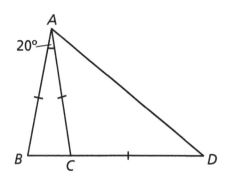

Not drawn accurately

Work out the size of angle *CDA*.

[3 marks]

Answer .... degrees

15 Here is some information about the colour of cars in a car park.

Colour	Frequency
Blue	7
Silver	8
Red	10
White	5
Green	6

Draw a fully labelled pie chart to show this information.

[4 marks]

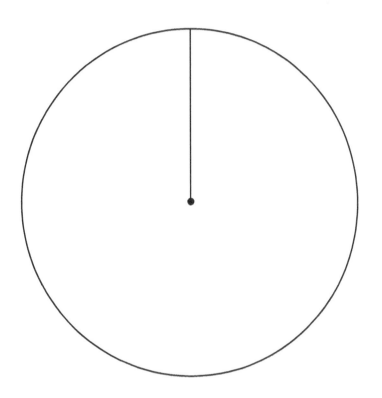

16 A cylinder has a base diameter of 20 cm and a height of 8 cm.

Calculate the volume of the cylinder.
Give your answer in terms of π.

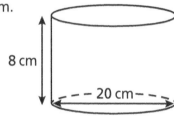

8 cm

20 cm

[2 marks]

Answer .... cm³

17 Solve $6x - 12 = x - 8$

[2 marks]

....

....

$x =$ ....

18 Work out the surface area of the cuboid shown.

[3 marks]

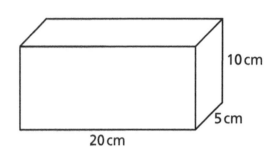

....

....

....

Answer .... cm²

19 Expand and simplify $4(x + 1) - 2(3x - 4)$

[3 marks]

Answer ....

20 The diagram shows one interior angle of a regular polygon.

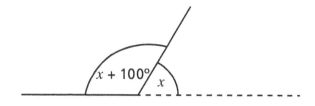

How many sides does the polygon have?

[3 marks]

Answer ....

21 The graph of $y = 2x^2 - 3x - 5$ is shown.

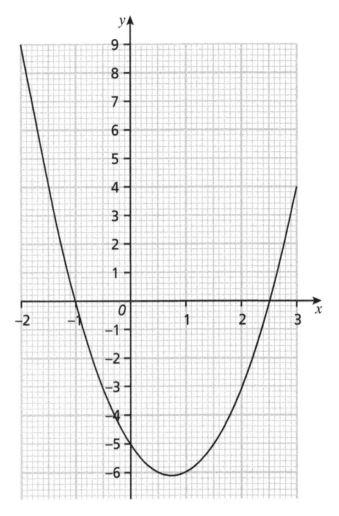

21 **(a)** Write down the values of x when $y = 4$

[2 marks]

Answer _____ and _____

21 **(b)** Write down the coordinates of the turning point.

[1 mark]

Answer (_____ , _____)

22 **(a)** Write 2.3×10^5 as an ordinary number.

[1 mark]

Answer ....

22 **(b)** Write 0.0005 in standard form.

[1 mark]

Answer ....

22 **(c)** Work out $2 \times 10^4 \times 8 \times 10^3$

Give your answer in standard form.

[2 marks]

Answer ....

23 Solve the inequality $3n + 7 > n - 4$

[3 marks]

Answer ....

24 Here is a right-angled triangle *ABC*.

Not drawn accurately

4 cm

x

B 6 cm C

Work out the exact length x.

[3 marks]

Answer .... cm

25 Here is a rectangle.

$3x$ cm

x cm

Not drawn accurately

The area is 48 cm²

Work out the perimeter.
You **must** show your working.

[4 marks]

Answer .... cm

END OF QUESTIONS

Collins

AQA

GCSE

Mathematics

F

SET B – Paper 2 Foundation Tier

Author: Keith Gordon

Materials

Time allowed: 1 hour 30 minutes

For this paper you must have:

- calculator
- mathematical instruments

Instructions

- Use black ink or black ball-point pen. Draw diagrams in pencil.
- Answer **all** questions.
- You must answer the questions in the space provided.
- In all calculations, show clearly how you work out your answer.

Information

- The marks for questions are shown in brackets.
- The maximum mark for this paper is 80.
- You may use additional paper, graph paper and tracing paper.

Name: _____

Answer **all** questions in the spaces provided.

1   Write down a number that is a multiple of both 5 and 8

[1 mark]

Answer ....

2   Increase 80 by 25%

[2 marks]

Answer ....

3   Work out the value of $3 + 4 \times 5^2$

[3 marks]

Answer ....

4   Here are four numbered cards.

| 5 | 7 | 6 | 4 |

4   **(a)**   Write down the biggest four-digit **odd** number that can be made with the cards.

[1 mark]

Answer ....

4   **(b)**   How many numbers between 4000 and 5000 can be made with the four cards?

[2 marks]

Answer ....

5   Here are six shapes.

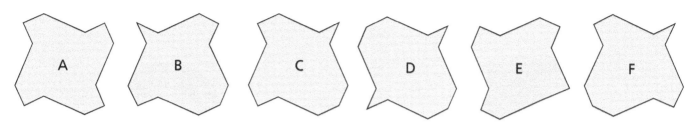

5   **(a)**   Which two shapes are congruent?

[1 mark]

Answer ....

5   **(b)**   Here is shape **A**.

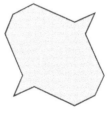

Write down the order of rotational symmetry of this shape.

[1 mark]

Answer ....

5   **(c)**   Here is shape **D**.

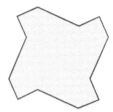

Write down the number of lines of symmetry of this shape.

[1 mark]

Answer ....

6   A shop sells beach toys.

Bucket £2.49   Spade £1.49   Rubber ring £3.50

6   **(a)**   Josie buys one of each item.

She pays with a £10 note.

How much change should she get?

[2 marks]

Answer £...

6   **(b)**   Josie receives her change in the least number of coins possible.

Which coins did she get?

[1 mark]

Answer ....

7 Round the number 278 to the nearest 10

[1 mark]

Answer _____

8 Here is a sequence.

5   9   13   17

8 **(a)** Write down the rule for continuing the sequence.

[1 mark]

Answer _____

8 **(b)** Write down the next **two** terms in the sequence.

[1 mark]

_____

Answer _____

8 **(c)** Work out the rule for the nth term in the sequence.

[2 marks]

_____

_____

Answer _____

9   A shape is drawn on a centimetre grid.

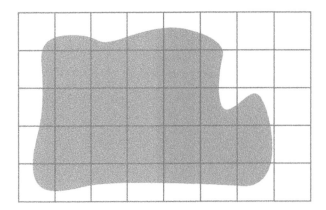

Show clearly that the area of the shape lies between 13 cm² and 33 cm²

[2 marks]

10   A running club meets each week.

The bar chart shows the number of runners that attend each week in January.

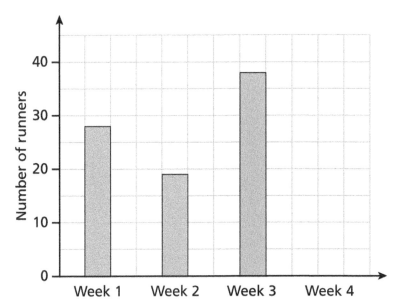

10 **(a)** How many runners attended in week 2?

[1 mark]

Answer _____

10 **(b)** How many more runners attended in week 3 than week 1?

[1 mark]

Answer _____

10 **(c)** The club has 60 runners.

55% of these attended in week 4.

Complete the bar chart.

[3 marks]

_____

_____

_____

10 **(d)** Jess said,

"On average, over half of our runners attended each week in January."

Is she correct?

Tick a box.

☐ Yes   ☐ No   ☐ Cannot tell

Give reasons for your answer.

[4 marks]

_____

_____

_____

11   On the centimetre grid below, draw

   a circle, radius 5 cm centred on *A*

   a 6 cm by 8 cm rectangle **inside** the circle

   a diagonal of the rectangle.

[3 marks]

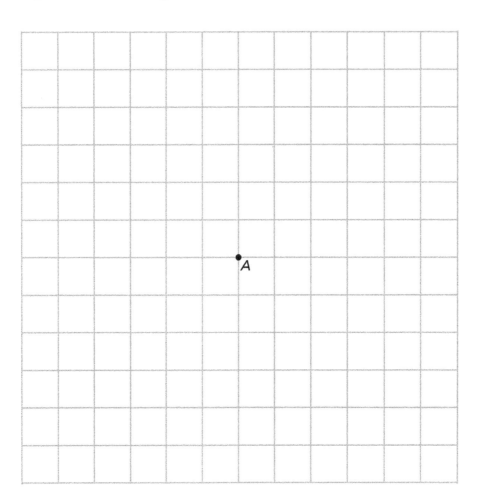

12 **(a)** Solve $\quad x - 9 = 17$

[1 mark]

$x = $ ....

12 **(b)** Solve $\quad \dfrac{x}{4} = 8$

[1 mark]

$x = $ ....

13 **(a)** Use your calculator to work out $\quad \sqrt[3]{46.656}$

[1 mark]

Answer ....

13 **(b)** Use your calculator to work out $\quad \sqrt{105} + 19.8^2$

[1 mark]

Answer ....

13 **(c)** Use estimation to show that your answer to part **(b)** is sensible.

[2 marks]

....

....

14   A bag contains 20 counters.

          8 of the counters are yellow.

          5 of the counters are blue.

          The rest of the counters are red.

      Work out the probability that a counter taken at random from the bag is red.

                                                                              [2 marks]

                                        Answer _____

15   This formula is used to work out the cost of a taxi fare.

      Fare = £4.00 + £2.25 for each mile + £0.75 for every minute stationary

15   (a) Jasmine takes a taxi that travels a distance of 7 miles and is stationary for 8 minutes.

           How much was her fare?

                                                                              [2 marks]

                                   Answer £ _____

15   (b) Alf takes a taxi that travels a distance of 6 miles.

           His fare is £21.25

           For how many minutes was the taxi stationary?

                                                                              [3 marks]

                               Answer _____ minutes

16 **(a)** Expand and simplify $(x - 2)(x + 3)$

[2 marks]

Answer

16 **(b)** Factorise fully $x^2 + 4x + 3$

[2 marks]

Answer

17 **(a)** Reflect the triangle in the line $y = -1$

[2 marks]

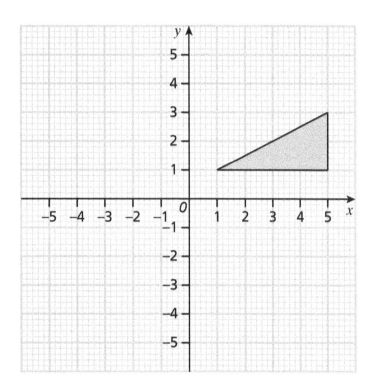

17 **(b)** Translate the triangle by $\begin{pmatrix} -3 \\ -4 \end{pmatrix}$

[2 marks]

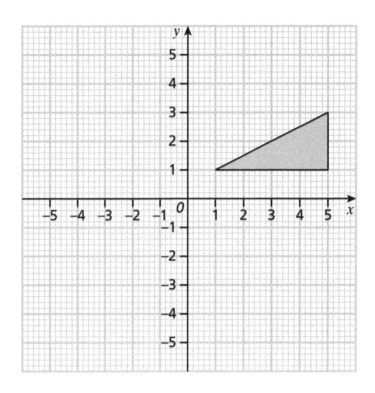

18 Work out the length x.

[3 marks]

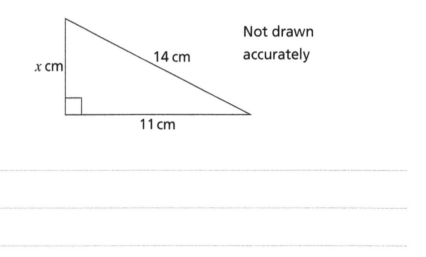

14 cm

Not drawn accurately

x cm

11 cm

$x =$ _____ cm

19 The table shows the heights of 40 young trees.

Height, h cm	Frequency	Mid value	
$140 \leqslant h < 150$	5		
$150 \leqslant h < 160$	9		
$160 \leqslant h < 170$	12		
$170 \leqslant h < 180$	8		
$180 \leqslant h < 190$	6		

Work out an estimate of the mean height.

[4 marks]

Answer _____ cm

20 **(a)** Work out 28 as a product of prime factors.

[2 marks]

Answer _____

20 **(b)** $20 = 2^2 \times 5$

Work out the lowest common multiple (LCM) of 20 and 28

[2 marks]

Answer _____

21   Triangles *ABC* and *PQR* are similar.

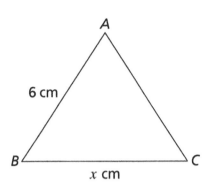 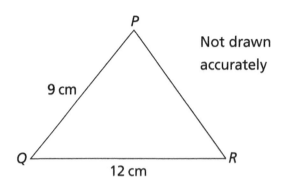

Not drawn accurately

Work out the value of x.

[3 marks]

Answer

22   A washing machine is reduced by 15% in a sale.

The sale price of the washing machine is £238

Work out the original price of the washing machine.

[3 marks]

Answer £

23 Two numbers are in the ratio 2 : 5

The difference between the numbers is 36

Work out the values of the two numbers.

[3 marks]

Answer .... and ....

24 The radius of this semicircle is 8 cm.

Work out the perimeter.
Give your answer to 1 decimal place.

[3 marks]

Answer .... cm

25 Using a ruler and compasses only, construct an angle of 60° at A.
You **must** show your construction arcs.

[2 marks]

A ―――――――――――――

END OF QUESTIONS

Collins

AQA
GCSE
Mathematics
F

SET B – Paper 3 Foundation Tier

Author: Keith Gordon

Materials

Time allowed: 1 hour 30 minutes

For this paper you must have:
- calculator
- mathematical instruments

Instructions

- Use black ink or black ball-point pen. Draw diagrams in pencil.
- Answer **all** questions.
- You must answer the questions in the space provided.
- In all calculations, show clearly how you work out your answer.

Information

- The marks for questions are shown in brackets.
- The maximum mark for this paper is 80.
- You may use additional paper, graph paper and tracing paper.

Name: _____

Answer **all** questions in the spaces provided.

1   This shape is drawn on a grid of squares.

1   **(a)**   How many lines of symmetry does the shape have?

[**1 mark**]

Answer ....

1   **(b)**   What is the order of rotational symmetry of the shape?

[**1 mark**]

Answer ....

2   **(a)**   Work out $4 \times \dfrac{2}{15}$

[**1 mark**]

Answer ....

2   **(b)**   Work out $12 + 3 \times 7$

[**2 marks**]

Answer ....

3   Write down **all** the factors of 20

[**2 marks**]

Answer ....

4 A shape made with 10 centimetre cubes is shown on the isometric grid.

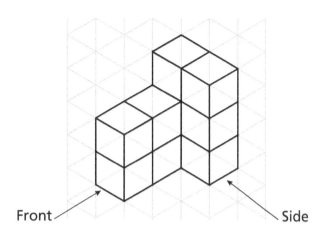

Front   Side

On the grids below draw the plan, side elevation and front elevation.

[3 marks]

Plan   Front elevation

Side elevation

5   A café has

a choice of three drinks   tea (T), coffee (c) or juice (J)

a choice of four sandwiches   salad (S), ham (H), meat (M) or vegan (V)

List **all** the possible combinations of a drink and a sandwich.
The first one has been done for you.

[2 marks]

TS

6   Here is a number machine.

6   **(a)**   Work out the output when the input is 5

[1 mark]

Answer

6   **(b)**   Work out the input when the output is 11

[2 marks]

Answer

7   This table shows the costs to use a swimming pool.

	Adult (16 and over)	Senior citizens (60 and over)	Child (5 years to 15 years)	Infant (Under 5)
Monday – Friday 6am to 9am 5pm to 8pm **Saturday** 6am to 8pm **Sunday** 8am to 2pm	£6.50	£5.00	£2.50	Free
Monday – Friday 9am to 5pm	£5.50	£4.00	£2.00	Free

7   **(a)** How much does it cost an 18-year-old to swim each day, Monday to Friday, from 8am to 9am?

[2 marks]

Answer £ ....

7   **(b)** A family go swimming on Friday at 1pm.

Their ages are

48   45   68   17   13   4

Work out the total cost.

[3 marks]

Answer £ ....

7   **(c)** A monthly pass costs £55 for an adult and allows the holder to swim at anytime.

Work out the most that an adult swimming 20 times a month could save.

State any assumptions you make.

[2 marks]

Answer £ ....

8   Here are some coins.

Tom and Jerry divide the coins.

The amount of money they now have is in the ratio 3 : 4

What coins do they each have now?   **[3 marks]**

Tom ....

Jerry ....

9   Here are eight numbers.

      3   8   6   9   11   12   5   2

Work out the mean of the numbers.   **[2 marks]**

Answer ....

10   Work out the size of angle x.

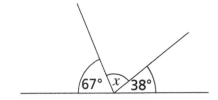

[2 marks]

$x =$ .... degrees

11 Large cuboids are 8 cm by 6 cm by 3 cm.

Small cuboids are 2 cm by 4 cm by 3 cm.

3 cm   6 cm   3 cm   4 cm
   8 cm   2 cm

11 **(a)** Show that the volume of **one** large cuboid is the same as the total volume of **six** small cuboids.

[2 marks]

11 **(b)** The large and small cuboids are stacked in alternate layers.

The bottom layer is one large cuboid.

The next layer is made from **six** small cuboids.

The total volume of the stack is 720 cm³

How many of each type of cuboid are used in the stack?

[3 marks]

Small cuboids ....

Large cuboids ....

12   Cereal is sold in two sizes.

A small box contains 350 grams and costs 79p

A large box contains 750 grams and costs £1.85

Which size is the better value?
You **must** show your working.

[3 marks]

Answer _____

13   The table shows information about three journeys.

Complete the table.

[3 marks]

Journey	Distance	Time	Average speed
A	32 km		64 km/h
B		1h 30 mins	50 km/h
C	50 km	50 mins	

14 A café owner records the average monthly temperature and monthly sales of ice cream over 10 months.

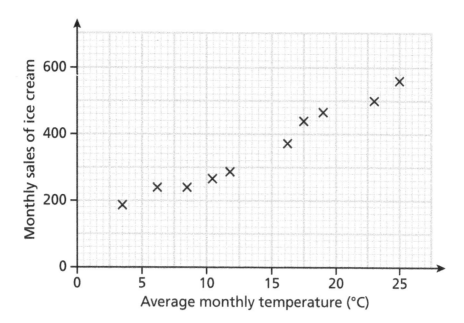

14 **(a)** Write down the relationship between average monthly temperature and monthly sales of ice cream.

[1 mark]

14 **(b)** The average monthly temperature for the next month is predicted to be 22°C.

Use the graph to estimate the sales of ice cream that month.
You **must** show your working.

[2 marks]

Answer

15 A two-digit prime number is one **more** than a square number.

Work out a possible value of the prime number.

[2 marks]

Answer ....

16 £3000 is invested in an account that pays 4% compound interest per year.

How much will be in the account after 2 years?

[3 marks]

Answer £ ....

17 (a) Simplify $x^3 \times x^6$

[1 mark]

Answer ....

17 (b) Simplify $x^{12} \div x^2$

[1 mark]

Answer ....

18 Here are two column vectors.

$$\mathbf{a} = \begin{pmatrix} 2 \\ 3 \end{pmatrix} \qquad \mathbf{b} = \begin{pmatrix} 6 \\ -2 \end{pmatrix}$$

Work out $2\mathbf{a} + \mathbf{b}$.

[2 marks]

Answer ....

19 Work out the next two terms of this quadratic sequence.

| 3 | 5 | 8 | 12 | 17 | 23 | ... | ... |

[2 marks]

Answer .... and ....

20 Enlarge the shape by a scale factor of $\frac{1}{3}$

[2 marks]

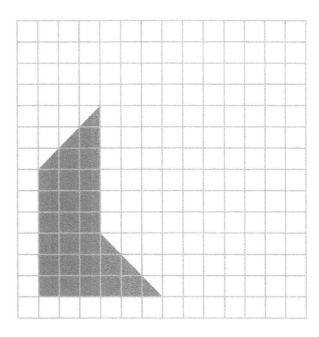

21 Two inequalities are shown.

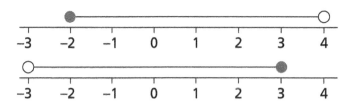

Write down the integers that are in **both** inequalities.

[2 marks]

Answer ....

22 Here are the equations of four lines.

Line A: $y = 3x - 4$ Line B: $y = 4x - 3$

Line C: $y = 3x + 3$ Line D: $y = -4x - 4$

22 **(a)** Which two lines are parallel?

[1 mark]

Answer .... and ....

22 **(b)** Which two lines intersect on the y-axis?

[1 mark]

Answer .... and ....

23 Match each graph to the equations.

Graph A   Graph B   Graph C

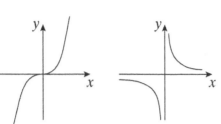
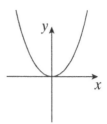

$y = x^2$ matches graph ....

$y = x^3$ matches graph ....

$y = \dfrac{1}{x}$ matches graph ....

24 Pressure = $\dfrac{\text{Force}}{\text{Area}}$

A force of 40 N acts over an area of 2.5 m²

Work out the pressure.

[2 marks]

Answer .... N/m²

25 Solve the simultaneous equations

$3x + 4y = 5$

$x + 4y = -1$

[3 marks]

$x =$ ....

$y =$ ....

26 A bag contains 10 balls.

4 of the balls are red and 6 are blue.

A ball is chosen at random and then replaced.

Another ball is chosen at random.

26 **(a)** Complete the tree diagram.

[1 mark]

1st ball   2nd ball

                                    Red

         $\frac{4}{10}$ Red

                                    Blue

                                    Red

              Blue

                                    Blue

26 **(b)** Work out the probability that both balls were the same colour.

[3 marks]

Answer

27   Write down the solutions to the equation $(x - 2)(x + 3) = 0$

[1 mark]

$x = $ .... or $x = $ ....

28   (a) Factorise $x^2 - 25$

[1 mark]

Answer ....

28   (b) Show that $(x + 2)^2 - (x + 1)^2 \equiv 2x + 3$

[3 marks]

....

....

....

....

29 **(a)** Show that the length x in the triangle below is 6.36 cm to 2 decimal places.

[2 marks]

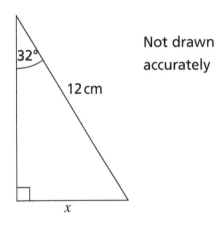

Not drawn accurately

12 cm

x

_____

_____

_____

29 **(b)** A cone has a half vertical angle of 32 degrees and a slant height l of 12 cm.

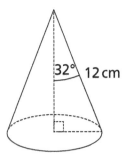

32° 12 cm

Work out the curved surface area of the cone.

The formula for the curved surface area of a cone is

Curved surface area = π × radius of base × slant height

[2 marks]

_____

_____

Answer _____ cm²

END OF QUESTIONS

Answers

Key to abbreviations used within the answers

M   method mark (e.g. M1 means 1 mark for method)

A   accuracy mark (e.g. A1 means 1 mark for accuracy)

B   independent marks that do not require method to be shown (e.g. B2 means 2 independent marks)

oe   or equivalent

ft   follow through

dep   dependent on previous mark

sc   special case

indep   independent

Set A – Paper 1

Question	Answer	Mark	Comments
1 (a)	1, 2, 4, 8	B2	B1 for three correct and none incorrect
1 (b)	e.g. 11 and 22	B1	Any two multiples of 11
2	16	B1	
3	$2x + 8$	B1	
4 (a)	Triangular prism	B1	
4 (b)	Six vertices **and** nine edges ticked	B1	With no other boxes ticked
5	$\dfrac{11}{4}$	B1	
6	125 : 75	M1	
	5 : 3	A1	
7 (a)	13	B1	
7 (b)	$3n$ seen Or show that the sequence is going up in 3s	M1	
	$3n + 1$ Or × 3 then + 1 implied	M1	
	151	A1	
8	64 '−15' then $\sqrt{}$ attempted	M1	In the correct order
	7	A1	Accept −7
9	40 ÷ 3 or 13 seen	M1	
	1.25 × 13	M1	Must attempt a partition method
	16.25	B1	
	£16.73	A1	scB1 for £17.50 (with no other method marks seen)
10 (a)	468 ÷ 4	M1	Accept ¼ of 468
	117	A1	
10 (b)	360° − 90° − 120° − 78° (= 72°)	M1	
	$\dfrac{72}{360} \times 100$	M1	oe
	20%	A1	

Question	Answer	Mark	Comments
11 (a)	$x = 15$	A1	
11 (b)	$3x \leqslant 4 + 5$	M1	
	$x \leqslant 3$	A1	scB1 for answer of 3 if M0
11 (c)	4 or x^3 seen	M1	
	$4x^3$	A1	
12 (a)	He has not subtracted the smallest value from the largest	B1	Accept 204 − 11 = 193
12 (b)	There is an outlier (11) which will affect the mean value	B1	
12 (c)	An attempt to order the data	M1	
	165 and 169 identified as the middle two numbers	B1	
	167	A1	
13 (a)	Answer in range 125°–130°	B1	
13 (b)	Answer in range 1.9 – 2.1 km	B1	
13 (c)	Bearing of 290° drawn	B1	
	Point D marked exactly 4.4 cm from T	B1	Point D must be on bearing of 290° for 2 marks
14	120 ÷ 5 × 2 (= 48)	M1	
	'120' − '48' (= 72)	M1dep	
	'72' ÷ [4 + 5] (= 8)	M1dep	
	40 minutes	A1	

Question	Answer	Mark	Comments
15 (a)	$50 \times (60 \div 6)$ (= 500) or $50 \div 6 \approx 8$	M1	
	$500 \times 8 = 4000$ or $500 \times 20 = 10\,000$ or $500 \times 24 = 12\,000$ '8' $\times 60 \times 8 =$ 480×8; 500×8 $= 4000$ or '8' $\times 60 \times 24 =$ 480×24; $500 \times$ $20 = 10\,000$ or '8' $\times 60 \times 24 =$ 480×24; $500 \times$ $25 = 12\,500$	A1	This answer mark will be affected by the assumption made in part (b)
15 (b)	An assumption supports their method in part (a) e.g. 'the machine operates for 8 hours per day' or 'the machine operates for 24 hours a day'	B1	
16 (a) Alt 1	$\dfrac{3}{25} > \dfrac{1}{10} > \dfrac{4}{50}$	B1	
	Bob	B1 dep	
16 (a) Alt 2	'Because they each did different numbers of trials'	B1	Accept similar statement
	'I can't tell'	B1 dep	Accept similar statement
16 (b)	Josh	B1	
	He did the most trials	B1	Accept similar statement
17 (a)	$\dfrac{6}{15} + \dfrac{5}{15}$	M1	oe, allow one error in the numerators
	$\dfrac{11}{15}$	A1	
17 (b)	$\dfrac{9 \times 1}{2 \times 6}$	M1	
	$\dfrac{3}{4}$	A1	oe

Question	Answer	Mark	Comments
18 (a)	30 seconds	B1	
18 (b)	$\dfrac{1200}{5}$ or $\dfrac{1.2}{5}$	M1	
	4 m/s	A1	
18 (c)	Straight line from (5, 1.2) to (6, 1.2)	B1	
	$\dfrac{1200}{6} \div 60$ (= 3.33... mins)	M1	
	Straight line drawn from (6, 1.2) to a point marked on the x axis between 9 and 9.5	A1	Point must be > 9
19	$4x$ or $x + 12$ seen	M1	Accept other letters used instead of 'x'
	$x + 4x = x + 12$	M1	
	3	A1	Trial and error scores zero unless final answer is correct
20	$320\pi \div 20$ (=16π)	M1	Allow $320 \div 20$
	$\sqrt{\,'16'\,}$ (= 4)	M1	
	'4' $\times 2$	M1	
	8 cm	A1	
21 (a)	110 000 000	B1	Accept 110 million
21 (b)	Five in the correct order	B1	
	France, UK, Egypt, Japan, US, China	B1	Fully correct
21 (c)	67 500 000 or 6.75×10^7	B1	In the 1st position
	68 500 000 or 6.85×10^7	B1	In the 2nd position
22	100 (mins) = 80% or (100 ÷ 80) × 100	M1	Allow any correct % equivalent e.g. 50 (mins) = 40%
	125 minutes	A1	
23 (a)	$\sin 30° = \dfrac{1}{2}$	B1	oe

Question	Answer	Mark	Comments
23 (b)	$\sin 30° = \dfrac{x}{5}$	M1	
	$\dfrac{1}{2} \times 5$	M1 ft	Allow ft from part (a)
	2.5 cm	A1	oe
24	$5^{-1} = \dfrac{1}{5}$ or $1 - 5^{-1}$	M1	
	$\dfrac{4}{5}$	A1	

Set A – Paper 2

Question	Answer	Mark	Comments
1	2.4	B1	
2 (a)	2340 g	B1	
2 (b)	640 cm	B1	
3 (a)	$2x - 2y$	B2	B1 for one correct term
4 (a)	$12a$	B1	
4 (b)	$12a = 60$ or perimeter = $6a + 8$	M1	
	$a = 5$ or $3a = 15$	A1	
	Perimeter = $15 + 4 + 15 + 4$ or $6 \times 5 + 8 = 38$ cm	A1	
5	$4 + 6 + 3 + 1 + 7$ ($= 21$)	M1	Allow 1 error
	3	A1	
6	$(-1, 2.5)$	B2	oe, 1 mark for each
7 (a)	$(36 \div 5) \times 3$ or $(36 \div 5) \times 2$	M1	
	£14.40	A1	
7 (b)	$\dfrac{16}{36}$ ($= 0.444...$) or $\dfrac{20}{36}$ ($= 0.555...$)	M1	
	$\dfrac{20}{36} \times 100$	M1	
	55.5...% or 55.6% or 56%	A1	Allow correct rounding of 2 significant figures or better
8 (a)	Primary **and** continuous	B1	With no other boxes ticked

Question	Answer	Mark	Comments
8 (b)	e.g. On average boys are taller	B1	One comparison using median
	e.g. Boys' heights are less varied	B1	One comparison using range
9 (a)	$2 + 12 - 8$ ($= 6$)	M1	
	No, it equals 6	A1	
9 (b)	$2 + 3 \times (4 - 8)$ or $2 + -12 = -10$	B1	
10	At least two prime factors found, e.g. $2250 = 2 \times 1125$ and $1125 = 5 \times 225$	M1	
	$a = 2$ and $b = 3$	A1	
11 (a)	13 in the eggs circle	B1	
	10 outside the circles	A1ft	Allow a ft mark for a correct answer leading from a correct method using their '13'
11 (b)	$\dfrac{'10'}{50}$	M1ft	Allow ft from part (a)
	$\dfrac{10}{50}$ or $\dfrac{1}{5}$	A1	oe
12	All numbers correctly converted to decimals or percentages e.g. 0.42..., 0.41, 0.385	M1	
	38.5%, 0.41, $\dfrac{3}{7}$	A1	
13	$\dfrac{3}{5} : 1$ or $\dfrac{6}{5}$ seen	M1	oe
	$\dfrac{3}{5} : 1 : \dfrac{6}{5}$	M1	oe
	$3 : 5 : 6$	A1	
14	20 in the centre	B1	
	1 and 20 as a factor pair	B1	
	2 and 10 as a factor pair	B1	

Question	Answer	Mark	Comments
15	2 × 100 × 100	M1	
	20 000 cm²	A1	
16 (a)	$4(x^2 - 3x)$ or $x(4x - 12)$	M1	
	$4x(x - 3)$	A1	
16 (b)	$4x < 18$ or $x < 4.5$	M1	
	$x = 4$	A1	
17 (a)	Angles in triangle are in ratio 2 : 2 : 1 or 5 parts altogether	M1	**Alternative method :** If $x = 72°$, $y = 36°$
	$y = 180 ÷ 5 = 36$	M1	Angles in triangle are 72°, 72° and 36°
	$x = 36 × 2 = 72$ degrees	A1	72° + 72° + 36° = 180°
17 (b)	72 degrees	B1	
	Alternate angles are equal	B1	
18	Equal arcs drawn from lighthouse and cliffs intersecting.	M1	
	A straight line drawn through the two intersection points	A1	
	A circle with radius 2.5 cm drawn from the yacht and the intersection with the perpendicular bisector clearly shown	B1	
19	1.5 × 0.5 (= 0.75 m²)	M1	Accept 150 × 50 = 7500 cm²
	$\dfrac{3000}{'0.75'}$	M1dep	
	4000 N/m²	A1	

Question	Answer	Mark	Comments
20	215 cm or 2.15 m seen or correct method to find m per worker e.g. 10.75 ÷ 5	M1	
	2.15 × 7	M1	
	15.05 m	A1	
21	8π	B1	
22	42, 84, 126, ... and 70, 140, 210, ...	M1	Allow errors if intention is clear
	210 identified	M1	Or a multiple of 210
	$x = 5$ and $y = 3$	A1	Or multiples of 5 and 3
23	Any translation	B1	The shape should be exactly the same size and orientation
	Fully correct translation Top right corner should be the point (4, 4)	B1	
24	Multiplier is 1.06	M1	oe
	£159 000 ÷ 1.06	M1	
	£150 000	A1	
25	Mid value completed 145, 155, 165, 175	B1	
	145 × 3 or 435 or 155 × 6 or 930 or 165 × 7 or 1155 or 175 × 4 or 700	M1	
	145 × 3 + 155 × 6 + 165 × 7 + 175 × 4 or 435 + 930 + 1155 + 700 or 3220	M1	
	3220 ÷ 20 = 161	A1	

Question	Answer	Mark	Comments
26 (a)	$y = \dfrac{5x}{3} + 1$	M1	
	<table><tr><td>x</td><td>–3</td><td>0</td><td>3</td></tr><tr><td>y</td><td>–4</td><td>1</td><td>6</td></tr></table>	M1	At least one of these points correctly plotted
	Fully correct line plotted	B1	
	$x = 1.5$, $y = 3.5$	A1	scB1 if correct answer with no graph drawn
26 (b)	$y = -x + c$	M1	Allow gradient = –1
	$x + y = 7$	A1	oe
27 (a)	9 pm	B1	
27 (b)	Falling at a steady rate	B1	
27 (c)	Line drawn from (2 am, 10) to (3 am, 13)	B1	Need not be straight but must be increasing

Set A – Paper 3

Question	Answer	Mark	Comments
1	9 tenths	B1	
2	–15	B1	
3 (a)	Isosceles	B1	
3 (b)	65 degrees	B1	
4	1.03, 1.3, 1.303, 1.33	B2	B1 for any three in correct order (ignoring fourth value)
5	36 – (11 + 10 + 7) [= 8]	M1	
	'8' ÷ 2 [= 4]	M1	
	Last two bars with heights of 4	A1	
6 (a)	48	A1	
6 (b)	No with 96 and 192 seen	A1	
7 (a)	(–2, 4)	B1	
7 (b)	7 identified as base of the rectangle	M1	Could be implied by correct diagram drawn
	(5, 4) and (5, 1) in either order	A1	Accept (–9, 4) and (–9, 1)
7 (c)	2 × 3 + 2 × 7	M1	
	20 cm	A1	

Question	Answer	Mark	Comments
8	Vertical line drawn up from H, then horizontal line drawn left from the top of the vertical line	M1	
	South East (SE)	A1	Allow correct bearing 135°
9 (a)	1.5	A1	
9 (b)	9.261	A1	
9 (c)	1024	A1	
10	Lists at least four of the factors of 40: 1, 2, 4, 5, 8, 10, 20, 40	M1	
	Lists cube numbers 1, 8, 27	M1	
	8 chosen	A1	
11	25 000 ÷ 2 = 12 500	A1	
	12 500 ÷ 84	M1	
	148.8…	A1	
	149 days	A1	
12 (a)	31	A1	
12 (b)	$\dfrac{\sqrt{59} + 5}{8}$	A1	
13	False	B1	
	True	B1	
	Sometimes true	B1	
	True	B1	
14	2 × 10.85 (= 21.70)	M1	A4 print is free
	21.70 – (2 × 3.09 + 1.52 + 3.80) [= 10.20]	M1	Allow 30.35 in place of 21.70
	2 × 5.95 – (2 × 1.07 + 3.80) [= 5.96]	M1	
	8.65 – (1.52 + 2.40) [= 4.73]	M1	
	'10.20' + '5.96' + '4.73'	M1dep	
	£20.89	A1	
15	£2500 × 0.03 × 2	B1	
16	51 30 13 / 17 / 21 15 / 6	M1	At least three out of six numbers correct
	Fully correct diagram	A1	

Question	Answer	Mark	Comments
17 (a)	584×0.188 [= 110]	M1	
	$\dfrac{584 - 312 - 110}{584}$ or $\dfrac{162}{584}$	M1dep	
	$\dfrac{81}{292}$	A1	
17 (b)	$312 + 30 - 12$ [= 330] or $584 + 11 + 30 + 6 - 10 - 12 - 7$ [= 602]	M1	
	$\dfrac{330}{602} \times 100$	M1dep	
	55%	A1	or better (54.817....)%
18	Complete method seen e.g. $\dfrac{19}{5} \times \dfrac{4}{3}$	M1	oe
	$\dfrac{76}{15}$	A1	
	$5\dfrac{1}{15}$ inches	B1	
19 (a)	At least eight points plotted correctly	B1	Allow ± 1sq accuracy
19 (b)	No correlation	B1	
	Correct interpretation e.g. 'there is no connection between height and salary earned'	B1	
20	4.8×10^8	B2	B1 for 480 000 000
21 (a)	€560	B1	Allow €550 to €560
21 (b)	Uses the graph to find 300 euros ≈ £270	M1	Allow £260 to £280
	'270' × 1990	M1dep	Converts any amount of £s to LBP
	Answer in the range (517 400 to 557 200) LBP	A1	

Question	Answer	Mark	Comments
22	Exterior angle = $180° - 150°$	M1	
	= 30°	A1	
	Number of sides = $360° \div 30° = 12$	B1	
23	Scale factor = $12 \div 8$ or 1.5 or $8 \div 12$ or $\dfrac{2}{3}$	M1	
	5×1.5 or $5 \div \dfrac{2}{3}$	M1	
	7.5 cm	A1	
24 (a)	$\dfrac{1}{3} \times \pi \times 3^2 \times 5$	M1	
	15π or 47.1... cm³	A1	
24 (b)	$336 \div 15\pi$ or $336 \div 47.1...$	M1	
	7.1... g/cm³	A1	
25 (a)	$(x \pm 3)(x \pm 2)$	M1	
	$(x + 3)(x - 2)$	A1	
25 (b)	0, −6 and −6 in the table	M1	
	At least six points plotted correctly from (−3, '0'), (−2, −4), (−1, '−6'), (0, '−6'), (1, −4), (2, 0), (3, 6)	M1	
	Fully correct graph joined with a smooth curve	A1	
26 (a)	0.2 on the 1st tail branch	B1	
	0.8, 0.2, 0.8 and 0.2 on the 2nd throw	B1	
26 (b)	0.8 × '0.8'	M1ft	
	0.64	A1	oe

Set B – Paper 1

Question	Answer	Mark	Comments
1	3500	B1	
2 (a)	7	B1	
2 (b)	7	B1	
3	Any fraction between $\frac{1}{3}$ and $\frac{3}{5}$	B1	e.g. $\frac{2}{5}, \frac{7}{15}, \frac{8}{15}, \frac{1}{2}$
4 (a)	4	B1	
4 (b)	11	B1	
4 (c)	4.25 + 2.75 + 1.5 or 8.5 or 17 × 2	M1	
	34	A1	
4 (d)	38 or 2 seen	M1	
	Half a circle drawn	A1	
5 (a)	985	B1	
5 (b)	167	B1	
5 (c)	138	B1	
5 (d)	32	B1	
6	2 and 5	B2	B1 for either answer and one wrong value, e.g. 2 and 7 B1 for both answer and one other value, e.g. 1, 2, 5
7 (a)	0724	B1	
7 (b)	(Arrives) 1005	B1	
	36 (minutes) + 1 (hour) + 5 (minutes)	M1	
	1 hour 41 minutes	A1	
7 (c)	0916 seen or 20 + 16	M1	
	36 minutes	A1	
8	Clear method shown (column, box, Chinese, partition)	M1	
	Correct partial calculation, e.g. 720, 48, 640, 128 or three out of four correct cells in box or Chinese methods	A1	
	768	A1	

Question	Answer	Mark	Comments
9 (a)	Four correct plots	B2	B1 for three correct plots or four plots with coordinates reversed
9 (b)	Parallelogram	B1	
9 (c)	Base is 6 and height is 4 or 6 × 4	M1	
	24 cm²	A1	
10 (a)	8a	B1	
10 (b)	6m or 30m	M1	
	36m	A1	
11 (a)	[37, 37.5]	B1	
11 (b)	40 (hectares)	M1	
	40 × 25 000	M1dep	
	£1 000 000	A1	
12	$\frac{4}{7} \times 56$ or $\frac{9}{11} \times 66$	M1	
	32 or 54	A1	
	86	A1	
13 (a)	Mark at $\frac{1}{3}$	B1	
13 (b)	Three odd and three even numbers	B1	e.g. 2, 3, 5, 6, 7, 8 is B2 2, 3, 4, 5, 6, 8 is B1 2, 3, 4, 5, 7, 8 is B1 2, 3, 5, 6, 7, 9 is B0
	Two multiples of 3	B1	
14	ABC or ACB = 80	M1	
	ACD = 100	M1dep	
	40	A1	
15	360 ÷ 36 = 10	M1	
	Angles calculated as 70, 80, 100, 50 and 60	M1dep	
	Angles accurately drawn	A1	
	Sectors labelled	A1	
16	$\pi \times 10^2 \times 8$	M1	
	800π	A1	
17	$6x - x = -8 + 12$ or $6x - x = 12 - 8$ or $5x = 4$	M1	
	$x = 0.8$	A1	oe

Question	Answer	Mark	Comments
18	Area of any face, e.g. 20×5 or 100	M1	
	$2 \times 100 + 2 \times 50 + 2 \times 200$	M1dep	
	700	A1	
19	$4x + 4 - 6x + 8$	M1	M1 for three terms correct
	$4x + 4 - 6x + 8$	A1	A1 for four terms correct
	$-2x + 12$	A1ft	ft on M1, e.g. $4x + 1 - 6x - 8 = -2x - 7$ is M1, A0, A1ft
20	$2x + 100 = 180$	M1	
	$360 \div 40$	M1dep	
	9	A1	
21 (a)	-1.5 and 3	B2	B1 each answer
21 (b)	$(0.75, -6.1)$	B1	
22 (a)	230 000	B1	
22 (b)	5×10^{-4}	B1	
22 (c)	1.6×10^8	B2	B1 for 16×10^7
23	$2n > -11$	M2	M1 for $2n > 3$ or $2n > -3$ or $4n > -11$
	$n > -5.5$	A1ft	ft on M1, e.g. $n > 1.5$
24	$x^2 = 6^2 + 4^2$ or $x^2 = 36 + 16$	M1	
	$x = \sqrt{6^2 + 4^2}$ or $x = \sqrt{36 + 16}$	M1dep	
	$x = \sqrt{52}$ cm	A1	
25	$x \times 3x = 48$ or $3x^2 = 48$	M1	
	$x^2 = 16$	M1dep	
	$x = 4$	A1	
	Perimeter $= 12 + 4 + 12 + 4 = 32\,$cm	A1	

Set B – Paper 2

Question	Answer	Mark	Comments
1	Any multiple of 40	B1	e.g. 40, 80, 120, …
2	Multiplier = 1.25 or 0.25×80 or 20 or 1.25×80	M1	
	100	A1	

Question	Answer	Mark	Comments
3	$3 + 4 \times 25$	M1	
	$3 + 100$	M1dep	
	103	A1	
4 (a)	7645	B1	
4 (b)	Any two numbers shown, e.g. 4675, 4657, etc.	M1	
	6	A1	
5 (a)	B and F	B1	
5 (b)	4	B1	
5 (c)	2	B1	
6 (a)	7.48 or 748 seen	M1	
	2.52	A1	
6 (b)	£2, 50p, 2p	B1ft	ft least number of coins for their answer for part (a)
7	280	B1	
8 (a)	Add 4 or + 4	B1	
8 (b)	21, 25	B1	
8 (c)	$4n + 1$	B2	B1 for $4n$ $(+ c)$
9	Marks on diagram showing counting of 13 whole squares within or 33 outside shape	M1	
	Explanation that area must be between these limits	A1	
10 (a)	19	B1	
10 (b)	10	B1	
10 (c)	0.55×60	M1	oe
	33	A1	
	Bar drawn to 33	A1	
10 (d)	$28 + 19 + 38$ + their week 4 or 118	M1	oe
	240 seen	B1	
	0.5×240 or 120	M1	
	Correct conclusion based on their total (No if correct)	A1	

Question	Answer	Mark	Comments
11		B3	B1 for circle B1 for rectangle (may be a different orientation) B1 for either diagonal (allow both drawn)
12 (a)	26	B1	
12 (b)	32	B1	
13 (a)	3.6	B1	
13 (b)	402.(2...)	B1	
13 (c)	Either value rounded to 1 sf e.g. 100 or 20	M1	
	10 + 400 = 410	A1	
14	$\dfrac{7}{20}$	B2	B1 for 7 seen
15 (a)	4 + 7 × 2.25 + 8 × 0.75	M1	Allow mixed units
	25.75	A1	
15 (b)	21.25 − 6 × 2.25 − 4 or 3.75	M1	Allow mixed units
	Their 3.75 ÷ 0.75	M1dep	
	5	A1	
16 (a)	$x^2 - 2x + 3x - 6$	M1	Four terms, with one in x^2, 2 in x and a constant term
	$x^2 + x - 6$	A1	
16 (b)	$(x + a)(x + b)$ where $ab = \pm 3$	M1	
	$(x + 1)(x + 3)$	A1	
17 (a)	Correct reflection, i.e. (1, 1) → (1, −3) and (5, 3) → (5, −5), etc.	B2	B1 for reflection in $x = -1$
17 (b)	Correct translation, i.e. (1, 1) → (−2, −3), (5, 3) → (2, −1) etc.	B2	B1 for correct translation of one vector component

Question	Answer	Mark	Comments
18	$x^2 = 14^2 - 11^2$ or $x^2 = 196 - 121$	M1	
	$x = \sqrt{14^2 - 11^2}$ or $x = \sqrt{196 - 121}$	M1dep	
	$x = \sqrt{75}$ or $x = 5\sqrt{3}$ or $x = 8.66...$ or 8.7	A1	
19	Mid value completed 145, 155, 165, 175, 185	B1	
	145 × 5 or 725 or 155 × 9 or 1395 or 165 × 12 or 1980 or 175 × 8 or 1400 or 185 × 6 or 1110	M1	
	145 × 5 + 155 × 9 + 165 × 12 + 175 × 8 + 185 × 6 or 725 + 1395 + 1980 + 1400 + 1110 or 6610	M1	
	6610 ÷ 40 = 165.25	A1	Accept 165
20 (a)	2, 2, 7 identified or 2 × 14 or 4 × 7	M1	
	2 × 2 × 7 or $2^2 \times 7$	A1	
20 (b)	2 × 2 × 5 × 7	M1	
	140	A1	

Question	Answer	Mark	Comments
21	Scale factor = $9 \div 6$ or 1.5 or Scale factor = $6 \div 9 = \frac{2}{3}$ or Ratio of side lengths = $9 \div 12$ or $\frac{3}{4}$ or Ratio of side lengths = $12 \div 9$ or $\frac{4}{3}$	B1	oe
	$12 \div 1.5$ or $12 \times \frac{2}{3}$ or $6 \div \frac{3}{4}$ or $6 \times \frac{4}{3}$	M1	
	8	A1	
22	0.85	B1	
	$238 \div 0.85$	M1	oe
	280	A1	
23	$36 \div 3$ or 12	M1	
	2×12 or 5×12	M1dep	
	24 and 60	A1	
24	Circumference = $\pi \times 16$ or arc length = $\pi \times 8$	M1	oe
	Perimeter = $8\pi + 16$ or 41.13…	A1	
	Perimeter 41.1 cm	B1ft	ft their perimeter
25	Arc from A cutting given ray	M1	
	Arc centred on intersection and crossing original arc plus line drawn	A1	Angle must be between [58, 62]

Set B – Paper 3

Question	Answer	Mark	Comments
1 (a)	0	B1	
1 (b)	2	B1	
2 (a)	$\frac{8}{15}$	B1	
2 (b)	$12 + 21$	M1	
	33	A1	

Question	Answer	Mark	Comments
3	1, 2, 4, 5 10, 20	B2	B1 for four or five factors
4	Plan Front elevation Side elevation	B3	B1 each Accept front and side elevation labelled the other way round
5	(TS), TH, TM, TV, CS, CH, CM, CV, JS, JH, JM, JV	B2	B1 for at least six listed
6 (a)	16	B1	
6 (b)	4	B2	B1 for 100 B1 for 0.4
7 (a)	5×6.50	M1	
	£32.50	A1	
7 (b)	$3 \times 5.50 + 1 \times 4.00 + 1 \times 2.00 + 0$	M1	oe
	$16.50 + 4 + 2$	M1dep	
	22.50	A1	
7 (c)	$20 \times 6.5 - 55$ or $130 - 55$ or assume pay £6.50 each time	M1	
	£75	A1	
8	$98 \div 7$ or 14	M1	
	42 or 56	A1	
	Tom 20, 10, 10, 2 Jerry 50, 5, 1	A1	Either order
9	$56 \div 8$	M1	
	7	A1	

Question	Answer	Mark	Comments
10	180 − 67 − 38	M1	
	75 degrees	A1	
11 (a)	3 × 8 × 6 or 144 or 3 × 2 × 4 or 24	M1	
	144 ÷ 24 (= 6)	A1	
11 (b)	720 ÷ 144 or 5 (layers)	M1	
	Small 12	A1	
	Large 3	A1	
12	350 ÷ 79 or 750 ÷ 185	M1	Allow mixed units
	4.43... or 4.05...	A1	
	small box	A1	
13	30 mins or 0.5 hours	B1	
	75 km	B1	
	60 km/h	B1	
14 (a)	More ice cream sold as temperature increases	B1	Accept positive correlation
14 (b)	Line of best fit	M1	
	480	A1ft	ft their line of best fit
15	17 or 37	B2	B1 for 26, 50, 65 or 82
16	1.04	B1	
	3000 × 1.04^2	M1	
	£3244.80	A1	
17 (a)	x^9	B1	
17 (b)	x^{10}	B1	
18	$\begin{pmatrix} 10 \\ 4 \end{pmatrix}$	B2	B1 for each component
19	30	B1	
	38	B1	
20		B2	B1 for any enlargement that reduces the size of the shape and keeps the sides in relative ratio. B1 for any three sides correct.

Question	Answer	Mark	Comments
21	−2, −1, 0, 1, 2, 3	B2	B1 for −3, −2, −1, 0, 1, 2, 3 B1 for −2, −1, 0, 1, 2, 3, 4
22 (a)	A and C	B1	
22 (b)	A and D	B1	
23	C A B	B2	B1 for one correct
24	40 ÷ 2.5	M1	
	16	A1	
25	$3x − x = 5 − (−1)$ or $2x = 6$	M1	
	$x = 3$	A1	
	$y = −1$	A1	
26 (a)	$\frac{4}{10}$ marked on red and $\frac{6}{10}$ marked on blue	B1	
26 (b)	$\frac{4}{10} \times \frac{4}{10}$ or $\frac{6}{10} \times \frac{6}{10}$	M1	
	$\frac{4}{10} \times \frac{4}{10}$ + $\frac{6}{10} \times \frac{6}{10}$	M1dep	
	0.52	A1	oe
27	$x = 2$ and −3	B1	
28 (a)	$(x + 5)(x − 5)$	B1	
28 (b)	$x^2 + 4x + 4$ or $x^2 + 2x + 1$	M1	$(x + 2 + x + 1)(x + 2 − (x + 1))$
	$x^2 + 4x + 4 − (x^2 + 2x + 1)$	M1dep	$(2x + 3)(1)$
	Shows subtraction of terms clearly	A1	
29 (a)	$\sin 32° = \frac{x}{12}$ or $12 × \sin 32°$	M1	
	$12 × \sin 32° = 6.359... = 6.36$ (2 dp)	A1	
29 (b)	$\pi × 6.36 × 12$	M1	
	[236.6, 240] cm^2	A1	

BLANK PAGE

BLANK PAGE